WHAT WAS WRONG?

EQUIPMENT MALFUNCTION IS NOT BAD, IF WE LEARN FROM IT

DR.R.JAYAPAL

Made with ♥ on the Notion Press Platform
www.notionpress.com

Contents

PREFACE

Majority of the industrial accidents or unwanted incidents can be traced to human negligence. Most of the time, managers do not need high grade technical skills but high-grade human understanding to resolve or prevent accidents.

This book gives a concise account of a few unwanted industrial accidents/incidents, written in story like form to find out root causes for a lighter reading but to provide a strong message. Indeed a technical story book. Generally, stories are interesting to read and keep the message in our minds for a long time. No special engineering knowledge is prerequisite to go through and appreciate this book. Just common sense is enough. However, Engineers and Scientists will find this book useful to apply in their work environment.

After my graduation in engineering from Indian Institute of Technology, Madras, India. I joined a reputed public sector company. After several years, I switched over to a private sector Company. In my more than three decades of experience, I had observed industrial accidents leading to production loss or life loss are mainly caused due to human negligence at one stage or other. Though we use sophisticated gadgets, software and algorithms to determine the root cause, most of the times, they are caused by human factors, which can be traced using common sense. But, how common is common sense? To instill interest among readers, I have explained some accidents/incidents I had come across, in story form for light reading. Nevertheless, any reader will get the strong message.

This book is intended for reading by all managers, managing any industry or business where there is more than one employee! No prior technical knowledge or experience is needed to use this book. It is a technical story book. All names/Characters are fictitious. Character Paul is my mouth piece!

Quality management professionals and teachers /students will find this book useful to fit these case studies in any of the root cause analysis tools to impart training or for learning. I am sure, after reading this book, you will care your employees and treat them as assets to lead accident-free industrial life!

ACKNOWLEDGEMENTS

Any accomplishment needs the support of several people and this work is of no exception. I thank my wife J. Leela whose support and encouragement were vital in completing this book. In my long career in industries, I had deep discussions with several engineers, supervisors, and workers who witnessed such accidents/incidents. I thank each one of them.I sincerely thank my parents,for their sacrifices to bring me up in my life.

I

Root cause analysis techniques–An overview

Even with a lot of precautions taken, accidents do happen in industry. Accidents are not wrong but if we do not learn lessons from them, it is wrong! For any accident, finding the root cause is a must. Root cause is nothing but the first event that triggered the series of events leading to the accident. We know the proverb "For want of nail, kingdom is lost". It means, in a battle during which the loss of a nail in a horseshoe led to loss of horse, which led to loss of the battle, which in turn led to lose of a whole kingdom. Needless to conclude that loss of nail is the root cause for the loss of kingdom, though exaggerated.

Root cause analysis (RCA) is a systematic process for identifying "root cause" of problems or incidents and an approach for responding to them. It is based on the basic idea that the management requires more than merely closing the problem that develop, but finding a way to prevent them. Root cause analysis helps to avoid treating the symptoms, rather than true underlying problem. RCA is an ongoing process that strives for continuous improvement. It means that achievement of total prevention by a single intervention is not always possible.

For effective RCA, following questions need to be answered

--- What happened?

--- How it happened?

--- Why it happened?

Basic methodology followed in most of the industries to determine the root causes are:

--- Define the problem

--- Gather information, data and evidence

--- Identify all issues and events that contributed to the problems or events

--- Discuss with all stake holders

--- Determine the root cause

--- Identify recommendation for eliminating or mitigating the reoccurrence of problems or events

--- Implement the identified solutions

Immediately after an accident, situation will be tense normally, lot of people giving a lot of reasons and solutions. First thing the manager has to do is to study the situation and console people who are affected. In countries like India, where union activities are high, union leaders are to be called immediately and discussed with, to satisfy their ego. Taking the affected to a good hospital etc. as per the need and involving their family members in attending to them are to be done on priority.

It is wise not to conduct root cause analysis at this stage. Manager should show at most concern over the affected people and impress the management to provide monetary and other support for the speedy recovery of the affected. If some machinery also got affected, priority is to be assigned to save the affected personnel rather than the machinery. If reverse is done, it will add fuel to the employees who are already hot, discussing the probable things that the management should have done to avoid such accidents. In such accidents, handling people is very much important. Any discussion by manager on loss of property, expenditure to management should be avoided till normalcy returns.

Most root causes can be traced back to the following attributes.

--- Competence of Personnel

--- Hiring qualified personnel

--- Lack of or insufficient training

--- Adequacy of technology or tools

--- Appropriateness of organization or department culture.

--- Health of the organization morale

--- Levels or number of resources (budget/personnel)

--- Decision making authority of the person or persons involved

Why determine root cause?

--- problems from recurring

--- Reduce rework and scrap

--- Reduce injury to personnel

--- Promote happy customers and stake holders

--- Increase competitiveness

--- Ultimately, reduce cost and save money

Though there are several techniques available, let us see some well-established techniques that are followed in industries for doing root cause analysis.

1. The "5-whys" Analysis

Toyota production system introduced this simple technique in 1970, in its factory. Strategy involves asking whys. Often the answer to the first „Why" prompts a second „why" and so on till real reason is obtained. In the case of "For want of nail, kingdom is lost", the following whys help.

--- Why?

--- Battle was lost (First level cause)

--- Why?

--- Loss of horse rider (Next level cause)

--- Why?

--- Loss of horse (Next level cause

--- Why?

--- Horse shoe lost (Next level cause)

--- why?

--- Loss of nail (Root cause)

Another simple example of five whys for root cause analysis of another problem

Problem – Flat tire or Car tire puncture

--- Why?

--- Nails on Garage floor

--- Why?

--- Box of Nails on shelf split open

--- Why?

--- Box became wet

--- Why?

--- There is a hole in the roof, through which rain water enters the garage

--- Why?

--- Some roof shingles are missing

Hence root cause for the flat tyre is missing of some shingles in roof. Cleaning the floor and fitting the missing shingles will solve the problem once for all.

2. Barrier Analysis:

It is a model used by some organizations to understand both why a problem happened and how it can be prevented. The premise of a barrier analysis is that a problem is prevented by having barriers in place to control hazards. There are three basic elements in barrier analysis; the target, the hazard and the barrier. The target is usually a person performing a job. The goal for personal safety is zero injuries. The target in the analysis is the person to be protected. Barriers provide control over the hazard by preventing it from reaching the target. When a barrier fails, the hazard reaches the target. Barriers currently in use in industries are mainly solutions from previous incidents.

3. Fish bone diagram/Ishikawa diagram

This tool provides a systematic way of looking at effects and the causes that contribute to those effects. This is also called cause and effect diagram, in view of the function of the fishbone diagram. The design of the diagram looks much like the skeleton of a fish hence the name fishbone diagram. Fish bone diagram is used to identify possible causes for a problem. Problem statement is effect. Various causes under broad headings like methods, machines (equipment), people (man power) materials, measurement and environment are discussed/ brain stormed. These are represented as horizontal line and arrows representing fish bone structure.

4. Pareto Analysis

This is a statistical technique in decision making, that is used for analysis of selected and limited number of tasks that produce significant overall effect. The premise is that 80% of problems are produced by a few critical causes (20%) hence this is also called 80/20 rule. Though surprising, following statements have been proved beyond doubt in majority of industries.

--- 80% of customer complaints arise from 20% of the products and services

--- 80% of delays in the schedule result from 20% of the possible causes of the delay

--- 20% of products and service in a factory account for 80% of profit

--- 20% of sales force produces 80% of company revenues

--- 20% of system defects cause 80% of its problems.

Quality department's job is to find out the 20% which contributes to 80% of problem and try to solve them. All the above methods are widely used in industries and in business houses. In situations where multiple causes can be attributed to a failure, majority use fishbone diagram approach. Intent of this book is not to teach the root cause analysis techniques in detail. Several books are available for this purpose. In this book, I have used why-why technique non-explicitly to find out the root cause of an accident/malfunction. Paul is a fictitious character and I have described the root cause, as if it has come from the intuition of Paul, to add spice.

For any industrial accident or malfunction involving loss of production, people, resources, root cause must be found out by 'hook or crook', applying the known techniques or by intuition etc. Then only solution can be found out, corrections done to avoid further recurrence. Accidents can happen but not finding the root cause should never happen.

II

Are protection systems Gods?

It was around 8 PM on a cool night. Boss called Paul, an experienced Electrical Engineer, over phone and informed that they needed to go to a thermal power plant, 200 miles away, the next day morning at 8. Paul was a bachelor boy at that time and all he needed was just 30 minutes to get ready, so he woke up at early morning 7.30! Boss came at 8AM and they started immediately with one oscilloscope, one digital multimeter and a few items like tools, wires etc. Matter was this. Their company had developed a power electronics unit which could operate the coal shaker. In thermal power plant, raw coal is taken through conveyor system and poured into gravimetric feeder through a hopper. It has to be shaken periodically to ensure slightly wet coal lumps also fall easily into the hopper. The power plant had placed a development order with their company. They had assured them that with the shaker attachment, coal flow could be smooth without clogging.

Station manager had given permission to test the equipment at 10 PM on that night. Technicians installed the hammers in the hopper and connected them with the power electronics unit. Boss as usual prayed god and switched on the equipment. Balakumar, company technician, found that phase and neutral pins were interchanged in the single -phase wall socket. He switched off the equipment and set the pins right. Boss offered second prayer and switched on again. Balakumar, suddenly switched off and said "I have to find out whether earth pin is proper". He did something and said earth was proper. Boss made third prayer and switched on the equipment. Equipment was required to operate the hammers in sequence. There was a heater circuit which worked fine but the hammers did not operate. Balakumar and Paul exhaustively tested everything, the circuit, the connections, component rating etc. By that time, it was 2 at night. Plant engineers became restless and told "you have to pack off at 5 AM". All were worried that a golden opportunity was getting wasted and only God knows when the next opportunity would arrive.

Boss suspected the electromagnetically operated hammers. It was Boss' idea to install hammers all- round the hopper and make them strike periodically one by one. Each hopper had four hammers. Several discussions took place in office whether to operate the hammers one by one or all at a time. Some even suggested to operate two hammers located diagonally opposite to operate at a time. Being microprocessor-based design, Paul had introduced all these options as selectable ones in the equipment. Hammers were disconnected from the equipment. A 24-volt DC power supply unit was brought and connected to one hammer. Balakumar switched on the power supply unit. Coil of the hammer got energized and was pulled away from hopper body. When this power was removed, coil got de-energized and the hammer which was pulled against a spring force hit the hopper. Similarly other three hammers were also checked. Boss got convinced that hammers are working. Balakumar was noting down the current taken by each hammer coil. Suddenly a spark came from Balakumar. He told Paul, "Sir, you asked me to increase the overload current setting in the electronic power supply. I forgot whether I did that".

"Oh God. That could be the problem", Paul said. Balakumar adjusted the potentiometer and increased the overload current limit setting.

All hammers were connected again. Equipment was switched on and switching sequence was commanded to start. Oh! All hammers operated well and the shaker started functioning. Plant engineer showed his happiness by patting Paul and Boss. He informed Plant manager and he came to plant at 4 AM, with sleepy eyes. He was happy to see the shaker operating well and coal flow was happening smooth as expected. Plant engineer had ordered tea and all sipped. Plant manager became very fresh and told Boss," I will order your equipment for all hoppers. Congratulations." Boss was very happy and hugged us.

As planned, they left the plant by 5 AM and went to hotel, slept for two hours. Boss has arranged a buffet breakfast in a star hotel. Boss asked Paul to explain the problem and solution. He said "Our equipment has current fold back type power supply protection. In this design there is a provision to limit the current so that it will not exceed this limit even in any fault condition. The current limit set was sufficient for normal operation. When the hammer operation was initiated, current exceeded the limit and DC voltage dropped. That was not sufficient for hammer operation. When Balakumar increased current limit setting, situation got changed. We should have checked this in our factory itself but got missed." Boss said "Oh! So much is there in this small protection circuit. Anyway, it is good that you people could solve this at least now."

True. Protection circuits are essential but at the same time they should not hinder the normal operation. If only they were attentive in this aspect, they would have saved at least five precious hours and unnecessary tension.

III

Can simple voltmeter help in complex plant control?

It was Bangalore, India. One reputed laboratory providing services to aerospace development wanted to build a high pressure and high temperature air delivery system. Paul's company got the order to design, manufacture, and supply and install the above facility. Company completed the manufacturing and their engineers were in the process of installation of all equipment. Client had specified conditions such that the air should be delivered at set pressure, set temperature giving the required flow. These parameters were so stringent, very elaborate control system had to be built up.

Several sensors and transmitters for pressure, temperature, flow measurements, several valves for their control were part of the control system. Distributed control system architecture was deployed. This control architecture was fairly new to India at that period. Paul's company worked hard to understand this new technology and executed the commercial order. It was a bold step. Generally, Indian companies are good at taking giant leaps in technology, even if it calls for very hard work and taking reasonable risk.

Operator could get the plant view in the computer screen and he could push a few buttons in the touch screen to operate the entire plant. Though this is order of the day in any plant design now, it was a marvel two decades back, when the plant was engineered.

John, I&C engineer was very busy in installing the transmitters, after doing in situ calibrations meticulously. Client laboratory had excellent canteen facility, serving exquisite south Indian dishes. Boss had given the freedom to eat anything in the canteen and his company would foot the bill. John and his team enjoyed the canteen foodies and happily put in their hard work equally in job too. Indirectly, Boss wanted that only. They were enjoying tea every hour, Boss had to tell them to drink water also in view of their health. Whenever Boss visited the facility, he used to enjoy vada and sambar in the canteen. (vada and sambar are dishes that are a rage in India!). Boss used to tell Paul often "Hey! If only I had put up a stall is a busy area and sold vada and tea, I would have made millions in a few years. In India, they are earning much more than engineers. You see, they do not pay income tax but you and I, bloody engineers, have to pay heavy income tax. "

What he told is true even now in India.

Coming to the main story, John completed the installation and switched on the panel. Paul and Boss were standing near the panel and watching. All transmitters were connected to DC power supply unit and the voltage was set at 24V DC as per design exactly. Air and gas compressors were started. Gas was burnt in gas burners to raise the temperature of compressed air.

Fuel burners were started to heat up. Plant did not work and DC power supply tripped. Plant was restarted. Again, DC power supply tripped. Client's side plant manager and his couple of engineers looked at our actions quietly. Boss became nervous. Since the plant was checked section by section thoroughly, Boss thought integral testing should go through smoothly. Plant was brought into manual mode of operation. Plant operation was checked section wise and it was working well. No tripping of power supply observed. It was a surprise to everybody why when all sections were connected, plant was tripping.

Paul asked John to show him the dc power supply distribution drawing. All transmitter and electro pneumatic controller were supplied 24Volt DC by a single cable. All equipment were taking power from that cable. Paul asked John to connect a two cores wire between farther most transmitter and the DC power supply. It was done instantly. I smiled at the Boss. He asked us to start the plant. OMG! Plant started and there was no tripping of DC power supply. Plant was working well with a few attempts of controller tuning; the air delivery was done exactly as per the contract terms. Plant manager and his engineers thoroughly checked the plant for 3 hours. John used to abscond for a few minutes in every hour. Everybody knew where he would have gone. They patted each other and the Plant manager expressed his happiness. While John started the 96 hours mandatory trial run, Paul went to Plant manager's cabin to sign a few papers.

All were returning to hotel by car. Boss asked Paul what was the problem and solution. Paul said "you know a long wire connects each and every transmitter. That means, due to voltage drop in wire, the transmitters in farthest end got only 14 volts which is not sufficient for operation. Power supply tripped due to low voltage. There are no meters to show voltage given to individual transmitter in our design".

Boss said "I see, what you did to fix it?" Then Paul said "I measured the voltage with a multimeter at each and every junction and found the voltage to be 14 volts at the last transmitter section. Few more measurements near the last transmitter showed 15 volts to 16volts, so I asked John to connect a wire from power supply unit to last transmitter to form a ring main." Boss said "Now I understand. Ring main improved the voltage at the farther transmitter. Good job done. Tell me what do you want?"

Paul said "Buffet dinner in star hotel to our team". Boss sanctioned immediately. In any plant, ensuring right amount of voltage, AC or DC to the individual equipment is essential first step.

True. Even a simple voltmeter can help setting right the complex plant control.

IV

Can a simple insulation tape checkmate a ghost?

John became jittery when his landlord asked him to vacate his rented house, as he wanted to give his house to his daughter, who was to come on transfer. John had no choice but to vacate. John was very happy in that house and many good things occurred to his family in that house. Another thing was that he paid very low rent as he was living there over seven years. Any other rental house would put a hole in his pocket.

Some interesting thing happened, during lunch time conversation in the office. One colleague told, "Hey Folks. There is one two bed room apartment vacant in our complex. People who come there for rent run away within two months". He laughed and further said "One police officer could live there only for 3 months. Just packed off." Paul asked him," can you narrate why all are leaving?"

"There is ghost in that house. It switches on the calling bell twice or thrice around midnight."

"Every day...."

"Yes. Every day. Get up and see. Nothing will be there. Ghost is just playing".

"One day, two days may be okay but entire family becomes nervous after three days. House owner brought several priests from many churches to drive away the ghost. All attempts failed".

"Nobody likes living with a ghost in the same house".

"Yeah... Today it is harmless but can cause havoc when it gets angry".

"Rent must be less".

"Yeah..., It is very less. Landlord will be happy if somebody at least occupies that house. He lives in first floor and he wants company".

"John will save a lot if he courageously occupies that house. We can ask the landlord to remove the calling bell".

"Oh... No! His priests had asked him not to remove the bell. Then the ghost will move to some other place in the house, which may be dangerous".

"John. Why don't you consider that house? You and your wife are very courageous. After all, you do not have faith in ghost". John thought for a while and said, "I will definitely move there. House is located in a posh area and the rent is damn cheap".

"Then how will you handle the ghost".

"I will place Bible near the calling bell switch and show my faith." John made a phone call to his wife. Finally, they decided to move to that house.

After a month or so, John and his wife moved to their new house. Landlord was happy and said "John, it is a harmless ghost. Every month I use to offer expensive hot drinks to it and then I give to the tenant. Let me continue the same". John was little bit happy with the offer of free hot drinks. John and his wife were awaiting the bell to ring during midnight. It rang at 2AM, 2.25 AM and 4 AM. They did not get perturbed as it was as expected. Every night, bell used to ring at various times in night. They became familiar with that sound and in subsequent nights, they even ignored. He had tied Bible book near the bell switch. Whenever the bell rang in the night, he and his wife used to read a few verses from Bible and then sleep. This arrangement was harmless and comfortable. Ghost never played any other mischief.

The next Sunday, Paul visited his house. Paul told John that he would inspect the bell, switch and wire. Somehow, he wanted to zero in on the culprit. John said "Sir, do not spoil my free drinks. I can tolerate ghost". John was confident that Paul would somehow find out the cause. If Ghost theory was disproved, house owner would increase the rent and stop free drinks too. So, he asked Paul not to do anything about this.

However, Paul was curious to know the real cause. They were looking for the time when house owner was away from home. Fortunately, house owner had gone to another town for some purchases. Paul opened the switch box. Suddenly two lizards came out from the switch box. In that box, one switch for calling bell and another switch for light were there. By close observation, it was found insulation in two wires coming out from calling bell switch had peeled off due to heat.

Paul told "John, put insulation tape around these wires, you should not get shock when you open the box and accidentally touch the wires".

John wrapped the insulation tape and smiled at Paul. John said "Will this solve my ghost problem?"

"Chances are there, let us see". Paul left the house at 9 PM after dinner.

Next day, John came running to see Paul and said," Sir, calling bell sound in the night disappeared. We expected the sound every moment but no sound till morning. Sir, ghost has really disappeared".

"Cool down John, ghost has gone on vacation. It will resume the bell when it comes back!"

By that time, Boss came to office and occupied his cabin. John went in and talked with the boss. As expected, Boss called me and said," Surprising!

How did you drive away the ghost? Tell me the problem and solution".

Paul explained, "I did not believe in the ghost theory. When we opened the switch box, we saw two lizards. That means they are living there. I guessed when the lizards move inside switch box during night, they may instantly short the open wires where insulation is peeled off. It shorts the switch and the bell rings. This may happen two or three times in night when the lizards play inside the box and short circuit. Now insulation tape is wound and a few holes found in the switch box were sealed. Lizards could not go inside. Even if they go, they cannot short the switch due to insulation tape".

Boss said "Brilliant but you have spoiled John's free drink".

John quipped "Sir, in the morning, first thing, I explained everything to landlord. He was extremely happy. He promised not to increase the rent and continue free drink".

True. Many a times, a big problem may have a small solution. A small insulation tape has solved the landlord from problem he faced several years! Imagine if same thing happens in a shop floor, entire shop floor would have been abandoned because of ghost fear!

V
A Bit of coin problem

There was a boy's hostel, near to Paul's home. Whenever he crossed the hostel, he used to enjoy whistles, shouting and loud laughter by the inmates. He used to feel that he lost every such thing in life. He felt life was a tension both at office and at home. He often stopped near the hostel, wait for a few minutes to rewind.

The other day, his friend Samy called him and said "Hey! I will come to your house in another one hour. We have to go to the hostel near your house"

Paul said "Splendid. Are we going for a dinner there? People say dosa is excellent there".

Dosa is a delicious dish famous in South India.

Samy said "Hostel Manager will give us dosa, don't worry. Before that, we have to finish a job."

Paul's boss runs another company with Samy as Chief electrical engineer. That company provides electrical wiring services to homes, colleges, hostels and small factories.

Samy came to Paul's home in his new car and they went to that hostel. Problem there was tripping of electrical circuits and they could not switch on power. They were surprised how the entire hostel could plunge into dark.

They discussed for a while as how to proceed. Samy said "First let us switch on light and fans section by section and isolate the faulty section." Paul said, "Agreed. Let us start with rooms in ground floor".

First section alone was powered up. All lights and fans worked. They repeated the same in other circuits one by one. When they switched on section 6, entire hostel circuit tripped. It became clear that problem was there in circuit 6 only. They switched off all lights and fans in section 6. Then they switched on all lights and fans up to section 5. Things were perfect.

Samy said "Now we will switch on lights in section 6 one by one." When they switched on one particular light, entire hostel tripped. Samy and his team checked the wires in the switch board leading to this light. Insulation was perfect and no short circuit observed.

Paul whispered something into Samy's ears. He did not believe but promised to give a try. By that time, several students gathered there and asked us to explain the problem. Hostel manager asked the students to disperse and allowed us to do our job. We closed the room. Electrician removed the bulb. A coin fell from the holder, Samy and Paul just smiled. Bulb was refitted. Section 6 was again switched on and found working. Samy switched on all sections. All bulbs and fans worked. Students made huge noise to appreciate them.

As agreed, Hostel Manager offered hot dosas and they left the hostel. It took almost two hours to complete the job. When Hostel Manager asked the reason, Paul said it was just a short circuit due to insulation failure and no sabotage suspected. They reached Paul's office. Boss was very busy with one tender. In his usual style, Boss asked Samy "Briefly tell me what was the problem and how you solved".

Samy said "Problem was in one light circuit in section 6. Some students had put one coin in the holder and fixed the bulb. Coin shorts phase and neutral. So, the circuit trips as soon as you switch on the light".

Boss said, "Brilliant, how did you get such a suspicion? It is almost impossible to locate this".

Samy showed his finger towards Paul and said it was what he whispered in his ears.

Paul said with a little hesitation, "Sir, to tell you the truth, I played the same mischief in my hostel days once when I quarreled with the Hostel Manager."

Everybody laughed.

Boss said "Why should the entire hostel trip?"

Samy said, "It is to blame the poor fuse coordination and MCB coordination. When we did the electrical wiring, we did it. But subsequently, it appears fuse wires were not properly selected and whatever available were used".

Boss immediately assigned the job of fuse and MCB coordination and replace as per that. Boss knew the trick of converting any calamity into opportunity! He would talk to the Hostel manager and get this small contract approved.

True, students can create intelligent problem, posing challenge. This incident exposed poor fuse coordination issues too. Same mischief can happen in a factory too. If any worker wants to take any supervisor to task, he can play any trick. Anything is solvable but one has to face loss of production for few hours. To avoid all such issues, the best solution is to keep cordial relationship with everyone in the factory.

VI

Can a small valve jeopardize the whole system?

By then, a few years passed. Paul got married and his family moved to an apartment. That apartment had 120 flats. Nice environment and nice people! That apartment had installed pressurized water pumping system for distribution of bore well. Water from bore well fell into an underground sump. Three pumps driven by three 5kw motors sucked water from underground sump. All three motors were driven by VFD (Variable Frequency Drive). Two pumps worked always and third one was kept as stand-by. Dedicated PLC (Programmable Logic Controller) panel was used to control the pumps automatically. Each pump outlet was connected to a common header through Non-Return Valve (NRV). In that design, overhead tank was avoided totally. Some water was stored in common header at 2.8 bar pressure. When the households consumed water, header pressure went down and the pumps worked to compensate. When the consumption was low; one pump operated at various speeds through VFD to establish the pressure at the header. When the consumption was large, one pump operated at full speed and the second pump at various speeds through VFD. This excellent system was supplied by a well-known German company. System was operating 24 x 7 without any failure for a long time.

One fine morning, there was a huge noise in pump room. Even for low consumption, both pumps operated at full speed continuously. Motor current shot up like anything. "Thud, thud" noise from pumps vibrated the pump room. Apartment Manager tried to bring the service engineer immediately but was not successful. Situation went on like this for a few days. As water supply was available to all flats, inmates also did not bother much.

One month passed like this and the power bill shot up by 50%. Finally, service engineer came and checked each and every component.

Service engineer "Nothing wrong in VFD panel, motors, and pumps. Why it works at full speed even for low consumption is surprising".

Apartment Manager, "What for is this NRV?"

Service engineer, "It allows water to go in one direction only. Water can go from pump to header. Header to pump reverse flow is not allowed by NRV".

Paul had seen NRV failures often in industries. He told service engineer to analyze what would happen if one or two NRV failed.

Service engineer's face became bright. All the while, he was thinking about failure of PLC logic, VFD drive etc. which were complicated and expensive to repair. Moreover, he had to bring control panel engineer available in head office located 400 miles away. Pointing suspicion towards NRV was comfortable to him. He started analyzing in that angle.

Service engineer, "yes. If one NRV fails, water can go to the inlet of next pump and there will be circulating effect. Now I almost got it. It is NRV failure. That is why even if water is not drawn by housemates, pumps are running for internal circulation."

Service engineer removed the NRV in first pump. It was okay. He fitted back and removed NRV in second pump. Flap was broken and it allowed water to flow in reverse direction also. Finally, problem got located. He arranged to bring one NRV from his office and fitted the same.

He switched on the pumps. "Thud, thud" sound completely stopped. Power consumption became normal. Service engineer left with a sense of satisfaction.

In a sophisticated control system, even a small valve can jeopardize the whole operation. In any installation, non-return valves are weak links. They can fail at any time. In any process industry, such valves are used in good numbers. One has to keep a watch on these valves. Doing "what if" analysis will help in analyzing any component failure.

VII

How serious is reading operating manual?

It is a well maintained 210 MW Thermal power plant in North India in private sector. It used to rank high in plant load factor, coal consumption efficiency etc. Plant is always kept neat and clean.

One day, Paul got a call from Mr. Ashok, General Manager of that plant. Boiler feed water pump with high voltage motor tripped very often and that was his complaint. Ashok said, "Plant has three feed water pumps. Two are operational and third one is stand by. Two pumps are operating well but third pump trips often. As long as two pumps operate, well, there is no problem. If anyone fails, we will be in soup".

Paul just joked, "Which soup? Vegan or non-vegan? You know, I am pure vegan"

Ashok, "Hey, do not cut joke. I am serious. What could be the problem?"

Paul asked him to send an email giving rating of motor, overload relay setting, temperature raise etc. I received all within half an hour. All were correct and he could not figure out the reason. He was about 500 miles away from the project site. Going to site just for this appeared to be a waste. He asked Mr. Ashok to know who was operating the pump.

Ashok, "A newly joined guy. He is very sincere and does exactly what I say. I am not suspecting him".

Paul asked Ashok to go to the starter panel and wait for his call.

Ashok, "Yes I will call you after going to the starter location".

Within fifteen minutes, Ashok reached the location and called him.

Paul said, "Please remove the wires coming from DCS panel for auto start and put the pump in manual start mode. Keep the phone in line and ask your staff to do that"

A few minutes passed Ashok, "yes, done"

Paul said, "Touch the rotor and tell me whether it is hot"

Ashok," Yes. It is hot."

Paul said, "Wait for 15 minutes and then start".

Ashok waited for 15 minutes and started the pump, after by- passing some interlocks in DCS. Ashok "Vow! Pump is now running smoothly. Let us run this pump for half-an-hour and report you." I was waiting for 45 minutes to get his call.

Ashok said, "No problem, pump is running well. Our boys say we will observe for two hours".

I said, "Ok. It is lunch time. Let me go home and have my lunch and a small nap. Nap of 15 minutes in sitting posture after lunch is recommended by heath specialists to feel fresh."

After two hours, Ashok phoned me and told that the pump is working very well and they have connected back to DCS.

Ashok, "I hope you will send me my one day's fee for consultation. Knowledge is never free"

Ashok, "Definitely"

Paul, "Did you read the operating manual of pump and motor carefully?"

Ashok, "No, as usual, people here know how to push the ON button and OFF button. That is all." Paul, "Okay. Okay. Don't be angry. As per standard, high voltage motor can be switched ON only three times in one hour, giving sufficient time gap. Your person has switched ON the motor, immediately after every trip. He would not have bothered about maximum number of starts permitted in an hour."

Ashok, "Oh I vaguely remember that commissioning engineer told me this point last year when this pump was commissioned. The new guy does not know this, it appears."

Paul, "It is clearly written in operation manual. Ask your guy to go through the manual thoroughly"

Ashok, "Okay. Okay I will instruct him immediately. But tell me, what is the reason for limiting the number of starts in HT motor?"

Paul, "Very simple. When you start the motor, huge starting current flows. Rotor gets heated up. Fan fixed with rotor is not able to sufficiently cool the motor."

Ashok, "Oh. Sufficient time is needed in between starts, to cool down the rotor."

Paul, "Yes. You got it. This is incorporated in the motor starter circuit. It will not allow you to start more than three times in an hour. Now ask your engineer to write a note to incorporate this restriction in operating manual."

Ashok, "Thank you. You solved my problem. We will meet again."

It is a lesson learnt, re-learnt by Mr. Ashok, that operating manual should be thoroughly read, before operating any equipment. It applies to any equipment, big or small. One can even conduct oral tests for operators asking questions only from operating manuals to test their knowledge, periodically.

VIII

How common is common sense?

It was more than thirty years when split ACs were getting introduced in Indian market. People used to debate which AC was better, split or window type. Paul's friend was running a small guest house in a small town. He wanted to buy split ACs for his guest house having 20 rooms. He identified one company which marketed split ACs in India. He tried for a deep discount. Company agreed to offer 40% discount, if transportation and installation were taken care by the buyer. Paul's friend readily agreed and placed order for 20 units. Paul was at Chennai for one week or so to attend a family function. Air conditioners arrived to the guest house and the technicians completed the installations in all rooms by reading instruction manual sent by the company. Upon completing the installation in all rooms, technicians found that air conditioners are not working properly. Rooms became hot instead of cool! Paul's friend read the instruction manual several times and found that the installation was done exactly as per the manual. He asked Paul to come to his guest house at Chennai to inspect. After his visit, he wanted to complain to the supplier asking for replacement.

Paul visited his guest house two days later. He went through the instruction manual thoroughly and inspected the installation. Fixing of compressor unit, cool air distributor and the metallic pipe connecting both the units were found to be exactly as per manual. Paul checked up the supply voltage for any abnormality. It was okay but the rooms became hotter and hotter as the Air conditioners ran.

Paul never saw any split Air conditioner till then. It was new to India at that time. He felt there was something basically wrong. He went to the terrace where distributors were located. To his surprise, they were gushing out chill air! Height of stupidity! Compressors were fixed in the rooms and distributors in terrace! Paul called the technicians and asked why they did like this. They explained compressor unit was big in size like window AC and hence fixed them in the rooms. Technicians said that there was no mention about location of compressor or distributor in the instruction manual.

Paul again read the manual. Technicians were right. Manual talked about how to fix compressor unit and how to fix distributor unit. It did not talk about where to fix these units. Manufacturer might have thought it was common sense to fix compressor unit outdoor and distributor indoor. But how common is common sense?

Paul asked the Technicians to fix one compressor unit in terrace and its distributor unit in room. They did it with a lot of reluctance as it was rework. Room became cool. Paul's friend felt bad for his ignorance and started blaming the instruction /installation manual for omitting such a vital point. Technicians refitted all the air-conditioners correctly and his friend heaved a sigh of relief.

Can an instruction manual assume a vital thing as common sense and leave, especially when the product is just being introduced in the market?

IX

When you analyze plant behavior, why do you leave human behavior?

It was a Steel plant. They use to manufacture seamless steel tubes for high pressure steam applications. Generally, pipes and tubes are manufactured from steel sheets. They are bent to form a cylinder and the joined ends are welded continuously by a method known as seam welding. This tube cannot be used for high pressure applications as the welded area is weak and it will give way.

High pressure tubes are made without seam welding. That is why they are called seamless tubes. Stainless steel rods are heated is a furnace. They are sent through hot pressing and drawing machines one by one. Pipe diameter gradually reduces and the pressing operation stops after reaching the required external diameter. In another machine, hot spear like tool, pierces at the center of the rod, so that hole at the required diameter is formed at the center. Drilling the rod at the center is done and tube is formed to the required size. Hot pipes are cooled and thus the seamless stainless tubes are manufactured. This explains the manufacturing process in a layman's language.

This Plant was imported from Germany. As a cost cutting machine, proprietor of the plant imported the machines without automatic operation system. Paul's company was assigned the job of building system for automatic operation. Till then, the plant was operated manually with six operators in one eight-hour shift. Paul's team studied the entire plant by noting down the details of all motors and other items. Paul and his team sat with the operators during the operation and prepared the process flow chart. They identified the sensors needed.

It was decided to have the entire operation using PLC (Programmable Logic Controller) and an Industrial PC for data logging and control.

After a gap of three months, existing motor control panels, operator push buttons/ displays were hooked to PLC and Industrial PC. HMI (Human Machine Interface) software was customized for this plant. Plant was taken for shutdown for two days and all the controls and operations were transferred to PLC and PC. Operators were well trained by Paul's team and they were very happy to operate the plant though PC. Many functions like monitoring various parameters periodically, operating various motors in sequence etc. were done automatically and they had to simply observe the plant operation. Owner of the plant threw a party, in which Paul and team also participated. It was a happy occasion to all of them.

After seeing the smooth operation for three months, management decided to reduce the number of operators. This sent shock waves to them. Operators cursed themselves for cooperating with Paul's team to get this automation done. Strategy of that management was to spread this as rumor and watch the reactions of the work force before taking a final call.

Operators slowly understood the ground reality and made them elves ready for any eventuality. Days were running. One day, the plant manager called Paul and told that the plant was tripping often in night shifts and it had to be restarted after every trip. This leads to complications and production loss.

Paul and team went to the plant, stayed there for a few days. They observed that all equipment were in good state. Even during their presence, plant tripped a few times in night shifts. Surprisingly no tripping during day shifts. This added to curiosity. Paul could not understand the nocturnal behavior of this plant! Paul studied the electrical supply for any transients that could affect the control system. He had no clue. He and his team only watched the plant more curiously, expecting any clue.

One night, a trip occurred. Paul's engineer was in the operator room, near the operating console. This happened at early morning 4 O" clock. Operator first stretched his hands backside to drive away sleep. His left hand touched the "Emergency Stop" push button inadvertently. Plant got tripped. Since it was an inadvertent action, he did not notice it. "OMG, I got it", the engineer screamed. Matter was conveyed to plant owner. Paul had emergency meeting with plant officials. Generally, emergency push button is used to be kept in any plant operation system. In case of emergency, operator presses that button to switch off the plant. Generally, two push buttons connected in series and kept separated by half meter distance, are used as emergency push buttons. During emergency, operators should press both buttons simultaneously. This simple but essential concept was somehow missed in the design. Paul expressed his concern to the plant owner and rectified this defect. He provided two push buttons and operators were asked to press both buttons simultaneously, in case of any emergency. After that, plant ran smoothly and did not face any trip. But for this small point, they would have spent a lot of money and time in replacing hardware and software, without knowing the root cause.

Plant owner came forward to train the excess operators in other functions and deployed them there. Anxiety among work men came to an end and plant ran smoothly.

The whole episode gave a lesson to look into human behavior also instead of suspecting hardware and software only.

X

Is electricity pure and clean as we think?

One day, a mini steel plant owner Mr. Kumar called Paul over phone and informed that his one auxiliary plant is tripping often. He decided to visit his plant and did so after two days.

In the mini steel plant, one auxiliary plant, Air Heater, produces hot air at a temperature of around 500^0C. Problem reported was in that plant. There, gaseous fuel is mixed with air and fired in combustion chamber. Hot combustion gas use to come out and heat the ceramic pebble bed. In next cycle, air is sent through hot pebble bed. Air gets heated up and comes out from the plant at high temperature. This hot air is used in the steel plant.

Paul checked up the plant log book. Plant trippings were recorded. Plant tripped randomly. None happened in some days and one or two times in a few days. No reason could be found out from the log book. Entire plant operation was being done automatically by a relay-based control system. Several discussions were held with plant personnel. Many a times, discussion with worker would help to identify the root cause. No worker could attribute any reason for this random tripping. As per Indian tradition, they even conducted puja (Religious offering) to ward off evil spirits.

We studied the relay logic circuit of the plant. There were more than 25 relays and a few timers operating the plant using solenoid valves, motorized valves etc. Circuit was perfect and no loose connections noticed. Upon discussions, it was decided to look into power supply side. Power analyzer equipment was brought from another factory and was connected to monitor mains supply. Brown out conditions like voltage sag, swell was continuously recorded in the analyzer. Paul left the plant and promised to return after two days.

He returned after two days. His team studied the voltage disturbance record. There were two brown outs. Mr. Kumar asked Paul, "What is brown out?"

He said," You know black out. If the power goes off completely, it is black out. But if voltage goes down for a small period, it is called sag and if voltage goes high, it is called swell. Sometimes voltage may be zero for a very short time. These are known as brown out. They are bad to electrical equipment". One brown out was voltage sag for duration of around 50 millisecond and another for 80 milliseconds. Trip log showed plant tripping immediately after theses electrical disturbances.

Paul said, "Kumar Sir. We are near to the solution. These voltage sags were the reason for tripping. In one case voltage had fallen down to 108VAC and in another case it had fallen to 125VAC. After that, normal supply voltage of 230VAC resumed. During this period, the energized relays could not hold and dropped. That is the reason for plant trip".

Mr. Kumar and his engineers were very happy to know the root cause of the problem. Mr. Kumar asked, "Any way, I am not electrical engineer. Can you explain why voltage sags occur? Power plant generator does not produce sags. They give smooth supply only. Am I correct?"

Paul explained, "Electrical voltage at the generator end is smooth as per the standard. Our plant is located several hundred miles away from the power generating station. We get power through transmission lines running for several hundred miles. Brown outs, short power interruptions/voltage dips occur due to short circuits, lighting strikes on overhead power lines and heavy load switching. Modern speed control devices used in factories like variable frequency drives also contribute to this. Duration of such faults is very small in the order of a few milliseconds. Though major electrical equipment like motors, lamps, heaters ignore this small duration disturbances but protection/control equipment using contactors, relays may trip during this sag voltages and jeopardize the plant operation. Electrically held contactors and relays that control the machinery, trip out from 5milli second to 20milli second after power is removed. Each short voltage dip now becomes a power failure for them and the plant needs to be restarted."

Mr. Kumar asked his engineers to discuss on the subject. Everybody accepted the root cause. Then the discussion on the solution part came.

Mr. Kumar, "Can we not do anything with incoming power supply"

Paul said," First solution can be to introduce Voltage-dip proofing inverter. It is designed to maintain the switch gear control voltage constant during voltage dips, effectively keeping the plant connected to normal power supply. In this approach, normal power comes to the control system from mains. During voltage sags, a static switch connects the load to the Inverter. During this period, your control panel gets power from inventor. Once sag vanishes, power is fed from mains".

Mr. Kumar said, "what is the second solution?"

Paul said, "Replace the entire panel with 24V DC operation contactors and relays. Relays can be latched relays. Normal relays will be on when you give power supply. They will be off when you remove power supply. In case of latched relays, a small pulse is given to one coil to switch on the relay. Relay is held in ON Position by spring and other mechanical arrangement. It can be made OFF by giving another short pulse in another coil. Since power supply is not holding the relays, there is no problem due to sags."

Engineers discussed and they were not in favor of redesigning the entire control panel. They preferred the voltage dip proofing inverter. They bought the inverter within a week and connected to the control panel.

Plant was observed for a month and not even a single trip occurred. Mr. Kumar thanked Paul and team for solving his headache. They all enjoyed the new experience and the dinner hosted by Mr. Kumar.

For a common man, electricity appears pure and clean. When we look deep, we see a lot of problems and impurity in it.

XI

Is bird in the hand worth two in laboratory?

There was a giant power plant equipment manufacturing company, located in southern part of India. They also use to do research to develop technologies for alternate ways of electricity production from coal. Several years back, they were experimenting with coal gas-based Magneto Hydro Dynamics power plant to produce power. They had a small 5MW power plant in their campus. They used to produce gas from coal which was used as fuel. Gas was called syngas. This gas was poisonous due to large percentage of carbon monoxide content. Plant provided adequate instruments, gadgets to take care of human safety aspects. Plant had good number of carbon monoxide monitors along the gas pipelines. Alarms used to be generated, if carbon monoxide leak concentration exceeds a preset safe level. Staff were given adequate training to take care, in case, any carbon monoxide leak noticed.

Gas produced was stored in a storage system called gas holder. This acted as buffer storage. Plant got continuous gas supply from gas plant. For any interruption, plant used to get gas from gas holder. Thus, power production was continuous. On plant shutdown days, gas holder used to be cleaned by operators. This process generally took five to six hours. Plant used to be under shutdown for eight hours in a month. Operators used to check carbon monoxide concentration in the gas holder by a peculiar method. Compressed air was passed into the gas holder to drive away residual syngas.

New safety manager Mr. Alexander joined the plant just then. Company appointed him based on his long experience in handling hazardous gases and gas plants. He asked the operators to explain the procedure they follow, before going into the gas holder for cleaning. They said they would pump compressed air for three hours to drive out residual gas packets. Then they would insert a caged bird (Generally Dove) into the gas holder using a long rope. They would scan the entire gas holder with the bird. If the bird was alive after the entire scanning operation, they used to declare the gas holder as safe zone to take up cleaning operation. Alexander was shocked to hear. He was a bird lover and was dead against such unscientific approach towards safety. Operators were adamant to follow their age old, proven practice. Alexander allowed them to proceed but was personally supervising the entire operation.

After the compressed air operation, they scanned the gas holder with the bird in cage. It was observed that the bird was dead. Compressed air was pushed in for another half an hour. They started the bird scanning operation with a new bird. This time the bird survived. Bird was let off. Next, operator brought another bird and did scanning to reconfirm that the bird was alive.

After confirming, they switched on green lamp to show that the zone was safe. They went inside the gas holder and started the cleaning operation and completed in about three hours. Alexander went to his cabin with a heavy heart. After that, he went to his boss and described the barbaric way of a safety procedure. Boss smiled and said "Why do you worry so much about killing a few birds? It is a proven procedure. Not even a single accident took place, till now". Alexander had a tough time to convince his boss to go for a proven scientific method than a cruel unscientific procedure. Alexander's proposal was to buy two numbers of portable carbon monoxide monitors with built in alarm facility. Gas holders should be scanned with these monitors. Alarms were to be set at twenty parts per million of carbon monoxide in gas concentration, above which they would give loud audible sounds. Boss asked Alexander to call for a meeting with all operators, maintenance crew, a few engineers, and managers and explain in detail.

Alexander called for a meeting and explained his proposal. Tension mounted among operators and they objected it tooth and nail. One operator screamed, "You are trying to eliminate us"

Alexander said, "Friends. This is the way other leading companies do. This is the procedure accepted and laid down in all safety standards. If we do not follow all safety procedure, external auditors will not allow us to operate that plant. To allay your fear, I will get down first into the gas plant and then you can follow me".

Operator were taken aback and finally agreed for using the carbon monoxide monitors. These monitors were ordered with a leading manufacturer. Paul helped Alexander to select the best and most reliable model, though expensive. As usual next shut down came. Operators conducted a special puja (A Religious ritual, to please gods) to protect them. Alexander also prayed Jesus and inserted the portable carbon monoxide monitor into gas holder and scanned the entire gas holder. Carbon monoxide concentration in certain pockets was over 100 ppm. Compressed air was again admitted, till the monitor showed less than 20 parts per million. Only this time, alarm was not there. Alexander asked one operator to scan the gas holder with another monitor. There was no alarm as carbon monoxide level was less than 20 parts per million. Alexander went inside the gas holder and was happily smiling. Then, operators climbed into the gas holder and completed the cleaning operation. Alexander was happy that he would save a few bird lives!

Time was running and the plant saw a few shutdowns. Operators became experts in using carbon monoxide portable monitors. Things were going smooth and Alexander stopped getting into the gas holder, as the operators gained confidence. Monitors were recalibrated in the laboratory before each shutdown.

Alexander was attending a budget meeting on one shut down day. Suddenly SOS call came from gas holder plant. News was that four operators went into the gas holder and one got swooned. All were brought up on an emergency operation and were taken to hospital. Alexander rushed to the hospital. He spoke to doctors and found out that all were safe. Operators and workers virtually scolded Alexander and vehemently shouted to withdraw carbon monoxide monitors and permit then to use birds. Alexander was calm and took a few operators and engineers to the laboratory to examine the carbon monoxide monitors. His boss was very curious. He called Paul to rush to the plant to study the situation and find out the root cause. He brought Instrumentation engineer also to check the equipment. Both monitors showed the correct reading during calibration operation. When the calibration gas with 100 parts per million carbon monoxide content was admitted, both monitors did not give alarm, though the alarm level was set at 20 parts per million. This was shocking to Alexander as well as other engineers. Instrumentation engineer checked the monitors and found out that the batteries were weak. When battery was weak, alarm operation was disconnected to save battery. What a stupid design! Though the safety instruction said that new batteries to be used before each shut down, laboratory people ignored that and allowed the old batteries. This was the basic mistake. Laboratory in charge was charge sheeted for this gross violation. Operators were explained this problem and they assured to check replacing the old batteries with new ones and the calibration. Paul called the carbon monoxide monitor supplier and explained their design draw back. Monitor Company Chief apologized for this and he flew to the power plant, the next day. He promised to change the design in such a way that the display would read "Low Battery" and would not show any reading. Monitor Company Chief understood the danger in his design and apologized for what all happened. He replaced the existing monitors with improved fail-safe design monitors. Instrumentation engineer thoroughly checked the new monitors with new and run down batteries. Operators were also convinced.

Is it not the duty of everybody in safety department to ensure all instructions are followed systematically before start of any operation? Why they missed replacing the run-down batteries? What is wrong with the feeling of operators that "A bird in the hand is worth two in the laboratory."

XII

Shocking truth in "No shock box"

During that period, Paul was working as visiting professor in an engineering college. That college was lacking in experience in areas like industrial consultancy, project work development etc. College management appointed Paul to improve the college in those lines. He enjoyed his job there, as it involved more practical aspects than theoretical aspects, in which the academicians were better.

One day there was a discussion in electrical machines laboratory on electric shocks at home. A few days before only, there was news about somebody died in bathroom due to electric shock from Geyser. One lecturer showed a news item in a local newspaper regarding a magic box which could prevent electric shock. Generally, Electricity Distribution companies advise ELCB (Earth Leak Circuit Beaker) to be installed in homes. This device monitors the incoming current in phase line and return current in neutral line. They are supposed to be equal. If there is a difference, it means some current is going to earth through a person, which is electric shock. If the difference in current is above a set value, the incoming power is disconnected and the person is saved. But the magic box advertised said that there would be no shock even if one touches the live wire for a long time. They could not believe their eyes. One Lecturer said "Professor, it is a wonder. If it is true, the inventor deserves a Nobel prize".

All Staff gathered in electrical laboratory and brain stormed as what he would have done. They all took two cups of strong coffee in between but could not get any clue. One lecturer said "Sir, shall we call him to our college and ask him to demonstrate his equipment." Paul agreed. He telephoned to the inventor cum manufacturer and requested him to do a demo in the college. That Company was just a startup, having tall ideas about their business growth, like any startup in general. After talking to the inventor cum manufacturer, Lecturer said "Sir, he wants a certificate from our college if we are satisfied, apart from orders". Paul immediately agreed. He conveyed this matter to the management. Their management owned a few manufacturing plants and trading business, apart from running that engineering college. One young director, who had his education in US, spotted a lot of potential in that product and was prepared to acquire that start up. Expectations went up and they called the manufacturer to be at the college at 10AM the next day along with his equipment.

Next day came. Mr. Rajan, inventor cum manufacturer came to the college, as promised. All assembled in board room, expecting Rajan to explain his product. Instead, Rajan requested all to come to the electrical laboratory, see the demo and then assemble in board room.

Mr. Rajan told very clearly that his magic box is meant for home loads like lights, fans, mixer grinders etc. and not for industrial load. His objective was to protect the people in the house from shocks. His equipment had three terminal inputs ABC for three phases (RYB in India) and a neutral and output terminals for ABC (RYB) and a neutral. He demonstrated 1KVA No - Shock box. He connected 3phase inputs ABC (RYB) and neutral. Much to our opposition, he removed his shoes and touched the A phase live terminal. There was no shock. He touched it for not less than one minute. We were taken aback. Then he touched A phase terminal with left hand and B Phase terminal with right hand. He held them for more than two minutes. No shock! He asked a few of us to touch him. A few bold lecturers and professors touched him. No shock to anybody. There were continuous claps and all shook hands with Mr. Rajan. Nobody could believe their eyes and ears. Entire demonstration was shown live through video conferencing to the young director. Rajan asked all for comments. He said "What more you want?"

Paul had a small suspicion at the bottom of his heart. He grew suspicious, the moment Rajan said that No shock equipment was only for home use. In most of the homes, there were no three phase loads like three phase motor etc. Only single-phase loads mostly. He asked Rajan, whether he could test the output in oscilloscope. Rajan readily agreed. One lecturer connected phase A and neutral to the oscilloscope and the waveform was fine. He repeated for phase B and C Nothing abnormal. Paul felt he was wrong again but somehow, his conscious told him that something was definitely wrong. He asked the lecturer to connect all three phases to the harmonic analyzer to observe the phase difference. Lecturer connected and screamed "Sir, all three supplies are in same phase and there is no phase difference".

Rajan shouted "No. Your equipment is wrong. Do not test with wrong equipment and undermine my invention".

Paul told Rajan to cool down and he connected one phase sequence indicator. It was not showing the presence of three phase. Mr. Rajan got annoyed and shouted at us, saying "You people do not appreciate the invention. No shock. But you are trying to figure out what is there inside the equipment. I will not allow any further test". He packed all his items with anger and left the laboratory. The young director was unhappy with Paul's action and asked him to come to his office next day and explain the lacuna if any or face the consequence. Paul took this as a challenge to prove the equipment wrong. A few staff members, who had doubt in the equipment, met separately and started brain storming. They were asked to go the black board, draw circuit diagram and explain. Nothing was convincing. One person suggested that it might contain the ELCB (Earth Leak Circuit Breaker) inside. If it were to be so, equipment would have tripped, the moment one touched the live wire. But there was no trip. Hence his idea was rejected.

Finally, one junior lecturer came to the black board. He said "I got the clue the moment he told that his equipment is for house hold loads only".

He proceeded further saying, "His equipment is nothing but an isolation transformer with neutral floating. That is why, when he touched the phase wire, there was no current path to ground and he did not get any shock. A bulb was connected between phase and neutral. It glowed.

Everybody felt it should be the answer. Somebody asked," Okay. But how come, he did not get shock when he touched A phase and B Phase (R and Y Phase)?"

Junior lecturer smiled and told "I have an answer for this too. He has connected one single phase isolation transformer in A phase. B phase and C phase inputs are left blank and not connected to the transformer. I presume he has connected the same A phase output to the output terminals marked Phase B and phase C. That is why, the three phase outputs did not show any phase difference". Another professor said, "You are right. When he touched A phase terminal and B phase terminal, he did not get shock, as both terminals are of A phase (simply shorted) and there is no voltage difference". All appreciated the wisdom of the young lecturer. Paul hugged him and said "Excellent analysis. Definitely I will talk to the young director about you". The next day, Paul met the young director and appraised him about the product. Being an electrical engineer himself, he understood quickly. He knew ungrounded system would not be acceptable to electrical authorities and they would not permit manufacturing this product. Ungrounded system is safety hazard. Paul told him that the myth in the equipment was decoded by a junior lecturer. Young director was very happy to know that and immediately sanctioned a cash award to him, in appreciation of his sharp brain. He told Paul "Sorry sir. I used some harsh words with you. Foolishly I pinned a lot of hope on this equipment". Paul was a bit relieved and the meeting was over.

How can anybody believe a tall claim, without testing and verifying it? That too in a safety related issue.

XIII

Is fear over new technology always based on ignorance?

In the engineering college, where Paul worked as visiting Professor for a brief period, management decided to install a solar power plant. He was given the responsibility of selecting the supplier for installation. He called for offers by advertising in local newspapers for a 50KW solar power plant. Offers were scrutinized and several rounds of discussions were held with the supplier. Finally, one party was selected based on technical suitability and price.

Out of two types of offers, viz Stand Alone and Line interactive, Paul and team decided to go in for line interactive type. In this type, expensive battery banks were not used. One can generate power and use it. If excess power was generated, it could be sent to the grid and sold. At that time, this technology was fairly new and was just entering into the market. Management was happy that they selected a state-of- art technology. The supplier installed the solar panels in the terrace above the electrical machines laboratory. They installed the line interactive inverter system panels in the substation premises. In one fine morning, Chief executive officer inaugurated the system and was put to use. System was designed such that it supplied up to 50KW of power and the remaining power requirement was met from grid. Grid is nothing but the lines through which electrical utility supply power to the user.

This was going on for several months. One fine day, utility announced a power cut and the college staff thought that they would get power from solar plant, which could be used for essential loads like lights and fans in class rooms and offices. To everybody's surprise, inverter went off and it was not giving power. That was the time, they thoroughly studied about Line interactive Inverter Technology. As per that technology, the inverter would work, if only there was power in grid. Inverter generated AC power taking DC input from solar array. Its output frequency and waveform were expected to be in synchronization with grid voltage. During shutdown, since there was no voltage in grid, inverter had stopped working.

People started doubting the wisdom of selecting such a system. Naturally, people want power, when there was no power from grid. Management arranged for a meeting with staff to explain the situation. Paul was asked to explain. He said," If you want power, during shut down, you have to go for a standalone system. Comparatively, this is expensive, since battery banks are to be bought. We cannot sell this power to utility. Line interactive systems are comparatively economical and we can sell the excess power to utility. That is why we chose this technology".

One staff asked, "Why line interactive system stops working when there is no power in grid?"

Paul said," Line interactive system senses the voltage, frequency and waveform of the grid voltage to generate power. If there is no power in the grid, its voltage is zero and hence the inverter does not work"

Someone asked." Why can't they design the Line interactive inverter to work even if power is not there in the grid?"

Paul replied, "Look. It is safety oriented. On the power shut down day, electricians may be working in the lines, substation etc. If Line interactive inverter supplies power, what will happen to them? Even if there is no power shut down, they may switch off a circuit to do maintenance. If we feed power from our equipment to that line, what will happen to them? Hence, as a safety measure, line interactive inverters are designed not to generate power when there is no power in the grid"

Staff got convinced and the meeting was concluded.

Then came summer vacation for six weeks. College and hostels were completely closed. A few security staff and maintenance department personnel used to come to college. Solar power plant was expected to produce power more than the college need. So, it should go to grid and utility should pay the college.

After the summer vacation, college reopened as usual. The next electricity bill came. Everyone had a shock, virtually. Electricity bill showed that we had to pay to utility. Accounts officer checked the bills before the summer vacation. New bill was slightly higher than the old bill. It only meant that solar power pumped into the grid was not accounted properly and the energy meter showed as consumption only. Nobody knew the reason.

Paul thought he would better call the solar plant supplier and discuss. Supplier representative said, "Sir, Which type of energy meter, you are having? Is it regular type or Net Meter?"

It rang the bell in Paul's mind. He forgot to apply for Net meter with the utility company. Ordinary energy meters assumed energy to flow from grid to the user. But Net meter would account energy flow from grid to user as positive and energy supplied by user to grid as negative. Meter finally would show the difference.

Paul said," Sorry. It is slip on my part. I have not applied for Net meter with utility".

Supplier representative said," OK. Please apply today itself. We will assist you by expediting the utility to install it fast." Paul paid the necessary fees and applied for Net meter. Utility installed Net meter in 10 days. College had to wait for semester holidays of two weeks duration to check the meter. Yes. The meter showed the generated energy during vacation, the consumed energy and the utility billed for the difference. Everybody heaved a sigh of relief.

For any new technology, there will be opposition. It is purely based on ignorance. Once necessary knowledge develops; new technology will be welcome with red carpet.

XIV

Can superstition solve an industrial problem?

It was an automobile parts manufacturing plant. Mainly they were manufacturing fasteners and small components needed by cars manufactured by a famous car company. Plant had lathes, milling and shaping machines, rapid prototyping machines and sophisticated tool room. A few years back, there were around 100 employees and the plant worked in three shifts. Company had huge order book and the plant worked in three shifts. The employees were compensated very well. Apart from regular salary, they were given various allowances like entertainment, children education, leave travel concession and were given gift coupons often. Management was so generous that they distributed a good portion of profit with employees.

Situation started showing bleakness, when the talk of electric vehicles started, replacing gas or diesel vehicles. Indian government set a target of replacing majority of petrol / diesel vehicles with battery operated ones. The car company, to whom this company was supplying parts, started revisiting their long-range plan. Accordingly, they introduced a plan to reduce the production of petrol/diesel cars year over year. The companies which were supplying the parts started feeling the heat and they synchronized their production plans with that of the car company. This company also pruned their work force strength and started operating the plant in two shifts, between 6 AM and 10 PM. Work force was unhappy that their paradise was getting lost. They understood the situation and co-operated with the management.

Management appointed consultants to study the manufacturing facilities and to advice how to turn the facility to manufacture parts for the electric cars. The car company was scouting for collaborators who could supply technology for electric cars. They were looking for technology providers who could provide technology for electric car, with minimum modification in the existing gas car. That way, they can pull on their ancillaries to meet their requirements, with very less changes. Situation was so fluidic that tension prevailed everywhere right from the car company to all its ancillaries. There are millions and millions of car mechanic shops, cutting across the country, catering to servicing the petrol/diesel cars. Even they started worrying about their future in view of this disruptive technology.

After so many discussions with work force, knowledgeable people in this area, management decided to run the show as it was with reduced man power and decided to cross the bridge when it comes. During this fluidic stage only, a strange thing happened. One day when workers came for the morning shift at 6 AM, they saw shop floor B was flooded with water. One tap in a washbasin, used by workers to clean their hands after work, was open fully. Shop floor had at least one inch height of water. They all felt that one worker might have forgotten to close the tap when he left at 10 PM. Supervisor of shop floor B got in touch with his counterpart in second shift and explained the situation. As he was living nearby, came immediately. He felt sorry for the negligence of his work man. It took an hour to clean and dry the shop floor. One-hour precious production time was lost. Next shift started at 2 PM. Around 10 PM, when second shift closed, supervisor personally inspected all the taps and ensured all taps are closed.

Then came the next day. Again, shop floor B was flooded with water. The same tap which was open yesterday was open today too. That tap was closed and cleaning, drying operation took place. Plant manager visited the shop floor and enquired. All were surprised to see the tap was open again.

Security people who did night shift were summoned and were asked whether they noticed movement of any mischief monger roaming in the night. They expressed that they were very alert during their shift and did not notice anything wrong. Management decided to leave it at that point and look after the business as usual.

On the next day the workforce entered shop floor Band noticed the same flooding and the same tap in open condition. Workers suspected mischief of some ghosts. Though majority of people there did not buy that theory, they did not have any other answer. Workers tried to convince the management to bring in the service of exorcist and to conduct religious ceremony to appease ghost. Management reluctantly agreed and asked the workers forum to decide on the steps to be taken and the person who will conduct the pujas to drive away the ghost. Workers went round the city to enquire about exorcists and they selected one who demanded exorbitant fee. Management agreed to pay the fees. He was brought to the factory immediately and he started his exercise throughout the night and left the factory at 5 AM. He told management that he had driven the ghost out of the factory.

Next day started. Tap was in close position and there was no flooding. Workers were happy that they could provide a good solution. Management wanted to wait for a few days before taking a conclusion. As usual, next day came. No flooding in shop floor B. Tap was it closed position. Everybody heaved a sigh of relief. A few days passed like this. Sunday was a closed holiday. On Monday, when the workers came to the shop floor on first shift, they had the shock of their lives. Same tap was in opened position and the shop floor was flooded with more water. Everybody was worried that their earnest attempt failed and they had come to a clueless state. At this stage, Managing Director called Paul to come to his office. He had organized a meeting with his engineers and supervisors. Paul was full of surprise and could not provide any clue. Suddenly one young engineer (Age may be around 22, who might have just joined after his graduation) got up and told that they should install a few cameras around water taps in shop floor and run them throughout night. He suggested that this should be a guarded secret. Paul immediately appreciated him for his brilliant idea. Four cameras and one DVR (Digital Video Recorder) were installed. To drive away attention of workers, a few more cameras, were installed in other shop floors also. Idea was that nobody should know the operation.

Then came the next day. As expected, the tap was open and shop floor was flooded. Supervisor explained the workers about the installation of CCTV camera. Suspense was building up. Everybody was eager to know the outcome of CCTV footage. That engineer and Paul saw the footage of camera that was put near the tap which was getting opened. Footage showed nothing till 3 AM from 10 PM. Around 3AM or so, a dark object moved near the tap. On zooming the picture, the object was identified as bandicoot (Big Size Rat). It made a few jumps and climbed the wash basin. It opened the tap and drank water! Took bath and happily got down and ran away. Bandicoot, which opened the tap, did not bother to close the tap! That portion of footage was copied separately and showed to all the people in the plant. Everybody felt relieved. Shop floor was full of laughter. Everybody appreciated the newly joined engineer for his brilliant idea. Through it was a general idea, the pity was, nobody suggested except the young engineer.

Plumber came and fixed one wheel valve at the bottom of washbasin. Water would come only if both valves are opened. Sure, bandicoot will not understand this! Next day, it was observed that bandicoot opened its favorite valve but water did not flow. After several attempts, it left. Problem was solved once for all !

Paul told the Managing Director that young minds alone can provide out of the box solutions, while experienced engineers provide solutions to the problems if they were similar to what they had already experienced. We need both types of people to run any organization successfully. Workers were regretting for suggesting a non- scientific solution which forced the management to go in that direction. They went and apologized with the Managing Director.

Can superstition solve a problem, which cannot be solved by science? Like medical doctor, if diagnosis is perfect, solution will follow simply.

XV

Can one weak link break the entire chain?

It was an electrical insulator manufacturing plant, situated in South India. A small sized plant with around fifty employees, mostly, workers. In electrical transmission lines, insulator acts as support used to attach transmission line to electrical pole. It supports weight of suspended wire without allowing flow of current through tower to ground. Different sizes of insulators are used to support lines having various high voltage levels. About 30 years back, when India was concentrating on providing electrical supply to every nook and corner of the country, insulators were in great demand. Almost all insulator manufacturing plants were buzzing with activities.

Paul got a call from MD of that plant. He said" For the past one week, many insulators are failing in high voltage test. Can you and your team come and study? I have sudden spurt of orders. I am afraid whether I will meet the commitment with such type of issues".

He insisted Paul and his team to start immediately. It takes just four hours to reach that place by car. Paul asked his two engineers to pack up and be ready to travel with him in 2 hours. They might have to stay in the plant for two days to study the plant.

They got into the car exactly at 2 PM and Paul drove the car. Insulator plant was a new thing for his engineers and he promised to explain the process during travel. It was a plant, manufacturing porcelain string insulators suitable for supporting 33 kilo volt line. Paul explained the process to his engineers.

"Raw materials like China Clay, Felt Spar, Quartz and Alumina are crushed in ball mill. Water is added and made into a slurry. In the Filter press, water from the slurry is filtered and cakes are formed"

"Why add water in ball mill and remove in filter press?"

"Okay, when we visit the plant, you will understand"

"Yes Sir"

"Cakes are transferred to pug mill through conveyer belt and pugs are formed"

"Oh. What is pug?"

"You watch the plant and answer me when we return."

"Okay"

"Pugs are dried in electric drier unit. CNC machines are used to shape the pugs as per design. Shaped pugs are again dried of moisture. They are then dipped into glaze."

"Why glaze? You will tell or you want us to tell?"

"Don't lose patience. Glazing is done so that external moisture does not stick to the insulator"

"I see"

"Now the pugs are called insulators.These insulators are fired after glazing in the Kiln."

"Is that enough or something more is there" "Don't you need metal fitting to hold the insulator to the pole? Specially made metal rods are firmly fixed in the central holes of insulator. These rods are firmly held with insulator with special cement using an injection process".

"Okay we understand. Still, we have three hours. We will study the literature given by you on this factory".

"Go ahead. I need to drive carefully from now on as the road is not good."

At around 6 PM, they reached the factory. MD took them to the Conference Hall and gave refreshment and coffee.

Paul called for the product manager and quality control manager. They came almost immediately along with production records and testing records. It was observed that Ball Mill for crushing the raw materials, filter press and pug mill are common for the whole plant. After the pug mill, the production line had been divided into four lines (A, B, C, D). Each line had dedicated CNC machine, glazing unit Kiln, and rod insertion machine. After this, all go to high voltage testing station, a common facility, QC engineers used to test each and every insulator by applying high voltage as per their standard. Insulator must withstand specified high voltage for a specified duration without drawing any current from high voltage source. If it draws some current, it means, insulator has failed and is short circuiting.

As per the inspection record, several insulator failures were observed only for the past ten days. That too in Line B. This was observed by his bright engineer. So, something was wrong in Line B.

"Sir, let us deeply observe line B", said his engineer.

Paul started seriously looking at Line B without informing anybody. Production was going on as usual. In Line B, nothing unusual noticed in machine shop, glazing unit and kiln. Paul and team waited near the operator who was operating the rod insertion machine and applying cement to the rod by injection process. When operators in Line A, Line C, Line D in rod insertion and injection process took at least one minute per insulator for this operation, the operator in Line B took only a few seconds and was looking very casual. This raised some doubt. His engineer went close to him and started observing his activities. Similar operators in other lines were densely packing the cement between the rod and the insulator central hole by carefully pressing the cement several times and ensuring no voids. Operator in Line B was simply filling the cement and pushing it to the next stage. They noted the product serial number and waited for a few hours to get the QC test result. As they expected, the chosen insulator had failed in HV test!

Paul asked his engineer," Can you now guess the reason?" Engineer explained with 1000 watts smile in his face," There were voids and cracks in the cementing process. High voltage breaks down occurred in the air gaps".

On further enquiry, it was found that the operator in Line B was a new recruit who had joined just one month back. Plant supervisor assigned him that job, as the regular operator resigned and left. Supervisor neither told him the importance of his work nor the adverse effect of not doing the job properly. He did not know even the complete process. He learnt his job through the operator who left on resignation for an hour. That's all! Paul explained the fault of Line B operator to the supervisor. Line B operator felt ashamed for not doing his job properly which increased the failure rate of insulators. He assured to learn quickly and promised to become good operator shortly.

Supervisor put that operator in training for ten days. Similar operators in other lines were asked to share the load. Experienced operator showed him the trick of the trade. Insertion of the rod straightly in the hole, pouring the cement in correct quantity, careful compaction to avoid voids and cracks etc. Paul explained the whole episode to the MD. He appreciated their efforts for quickly resolving the issue. Paul said, "I owe this finding to my engineers." They stayed in the plant, the next day also and monitored the inspection reports. Failure rate had drastically reduced. It was up to the acceptable level. MD asked them whether they could help to make zero rejection. Paul assured to take up that issue as his next project. They bid good bye to MD and team and started their return journey. Engineer, while travelling said," Sir, shall I explain what is pug?" They all laughed and closed the chapter.

Job should be explained clearly to any new operator. He should be explained the importance of his work and how it can affect or effect the productivity and profitability of the company. He should be made to feel this. No job is small in a line assembly. Any weak link can cut the entire chain. If only the supervisor had explained the importance of his job and given training in the whole plant to understand the importance of his job in the whole process, costly rejections could have been avoided.

XVI

Do workers only err or even the bosses?

In southern part of India, one big company took a contract to design, manufacture, supply and commissioning of a large sized thermal power plant. It was a prestigious order, won against international competition. Plant was commissioned in record time and the plant performance was exemplary.

Just to tell in a nutshell, in a thermal power plant, coal is powdered in mills and sent to coal burners in the boiler for burning. This heat raises the temperature of water to make steam. This steam rotates a turbine. Generator, which is coupled to the turbine rotates and produces power. Temperatures and pressures encountered in the boiler equipment are very high, needing specialized design to ensure safety. Temperature in the boiler furnace can go to 1000^{0}C. Steam temperature goes up to 540^{0}C. With careful safety measures, power plant operates with minimum or nil accidents.

Paul's friend Sundar worked in that company as Chief of the Technical Services department. That department's role was to create instructions for safe and successful erection and commissioning of the plant. Being the first of its kind in that category for the company, Technical Services department worked hard to act as brain for site readiness, erection and commissioning of the plant. It trained the plant managers for safe and efficient operation. Sundar was in cloud nine. Paul congratulated him several times, as and when, reports came about smooth operation of the plant. Several customers, who wanted to put up similar power plants, visited that plant.

Paul, "Excellent Sundar I am proud of being your friend".

Sundar, "I am coming to the plant next Monday and plan to be there for a week to inspect all the components. I will be staying in company Guest house. I want you to come there and be with me. I know, you are also a boiler expert and our discussions will be really fruitful".

Paul, "No Sundar. I am not an expert like you but know something about power plant. Okay. I will definitely come and see the plant, you have designed.

Sundar, "Thanks a lot. I will make all arrangements for you. Bye".

Conversation ended. Paul reached the plant as per his schedule and stayed in Guest House along with him. They visited the plant inch by inch. They started from coal yard where coal was unloaded and stored. Huge conveyer arrangement took tons and tons of coal to the mills which powdered the coal and fed to coal burners in boiler furnace. These equipments were operated automatically, controlled by engineers sitting in central control room.

Sundar, "See how nicely we have organized the coal handling system. Ash handling system is also very elaborate".

True. Indian coal contains around 40% ash. Sundar used to jokingly comment that Indian power plants produce ash and power was byproduct! It is indeed a challenge to Indian engineers to produce power from high ash laden Indian coal.

After miles of walking, they felt tired and left for Guesthouse in late evening. They planned to visit the furnace area, the next day. Furnace height was more than sixty meters. Coal burners were kept at various levels. Stainless tubes carrying water go around the furnace to pick up heat. They felt the next day would be very hectic, without knowing something serious would happen.

The next day, they went to the plant as usual at 9 AM. Boiler maintenance engineers accompanied them. They wore helmets, fire proof aprons and safety shoes. There were walkways made of steel grilled structures, to go around the boiler. Staircases used to be there to climb from one level to next level. These levels were called elevations. Sundar whispered to me," Generally ladies are not allowed to go around the boiler in these elevations. Do you know why?"

Paul," Surprising. Will they keep quiet, if you declare so?

Sundar – "Now we are in fourth elevation. Look through the grill and see first elevation. Imagine what will you see, if there is a girl in first elevation! You will see her from the top"

Me," Naughty. Are you not old?"

Sundar, Yeah! "One is safety and another is to escape the sight from such boys"

They laughed for a while and climbed up. Weather was hot and humid. Sundar felt uneasy with his apron. He just removed it and held in his hands. They were climbing the stairs. Suddenly Sundar wanted to see the flame in the boiler furnace. At various levels, manufacturer used to provide small doors in furnace area. They are called man holes. Sundar was about to open a man- hole. Paul cried," Hey! Put on your apron and open slowly. Sundar, "I am a boiler expert. You are teaching me about this. Do you know that boiler furnace is always kept under negative pressure, Hot air cannot come out. If at all, some external air can go into the furnace"

Paul,"OK Sundar. It is theory. Please take safety measure. Hell with your expertise."

Boiler maintenance engineers also cautioned Sunder to be careful while opening furnace door.

Sundar, "I have seen hundreds of boilers. I know what I am doing".

XVII

Is proper storage of safety equipment while not in use, is important?

It was a day in 2010. When Paul was glancing through a newspaper, he read a news about an accident in one leading bearing manufacturing plant. Though no casualty reported, newspaper coverage of the accident was interesting. It read that one small gas cylinder flew like a rocket, made several rounds in the shop floor hitting a few people and some objects. He read the news item several times but could not know the reason for accident. He called his wife and showed the news coverage. She also laughed. How can a cylinder fly like a rocket? As she became more inquisitive, she asked him to find out the reason. He presumed that she might be internally worried about her cooking gas cylinder flying and making damages.

The next day, when he raised that topic in his office, everyone was surprised but no one could figure out the reason. Paul thought of catching hold of some people in that company and knowing the whole episode. With this intention, he surrendered the universal Guru Google. He went to company's website and searched for the list of important people. He could locate one person by name Roshan, who was the chief of maintenance department. Paul saw his photograph. It sounded a bell in him. After some recollection, he identified him as his B.Tech., class mate. If it was true, his job would be easier. His secretary connected his number. After a minute of talk, he identified him as his class mate, through his voice, photo and name. He recollected immediately and they spent a few minutes, going down the memory lane. Paul introduced himself as management consultant, well versed in resolving quality, production issues and good in doing root cause analysis to find out the genesis of any industrial accident. Roshan expressed his happiness and asked him to visit to his plant to help him in root cause analysis of that incident. Coincidently, Paul heard a small boy in TV singing a song "For want of nail, kingdom is lost". This was true in several industrial accidents. If we drill deep, reason could be attributed to a very small thing, we would not have suspected even in our dreams.

Paul visited that company, the next week, along with one of his engineers. Roshan took them to the plant. They made a general visit. Then, he took them to the shop floor, where that accident took place. Paul's engineer, quietly, disconnected from him and merged with the maintenance staff there. Paul knew he would talk to the correct set of people and collect information. In such industrial accidents, workers who were in the accident spot could give exact information which would lead to correct root cause. Many a times Managers' versions are either mutilated or exaggerated to add spice.

Plant had excellent order book. It worked in all three shifts and could just meet the delivery schedule. Plant used to operate even on Sundays. Company had declared 15 days spread over a year as holidays and plant used to stop production in those days only. Maintenance department would become very active during those days. They used to collect all repair works list and execute them during those days.

It was one such holiday. Instead of 3000 and odd employees, only 100 and odd maintenance staff had clocked their attendance. Welding work was going on in that shop floor. Carbon di-oxide based fire extinguishers used to be kept near the work spot to extinguish fire, if breaks out. To everybody's surprise, one small carbon dioxide cylinder started flying like rocket. It made a few rounds, hitting an object, changing direction, hitting a few more objects and people. Finally, its energy got drained and fell on the ground. This was exactly what happened on that day, as per Paul's engineer.

Workmen were well trained to use CO_2 fire extinguishers. The main danger with Co_2 extinguishers is that they will cause problems in a confined space. They starve a fire of oxygen but in a confined space, they will also reduce the amount of oxygen available to breathe. CO_2 extinguishers work by smothering the fire and cutting off the supply of air. Such fire extinguishers are fitted with frost-free horns, as the hand holding the horn can otherwise be frozen to the horn, as the gas gets very cold during discharge. CO_2 extinguishers are filled with liquefied CO_2. Normally cylinder pressure is 55 bar or 825 PSI. When the extinguishers is operated, the liquid expands to gas in the horn and get released into air neutralizing the oxygen that the fire is feeding on, disabling the fire's ability to spread. His engineer collected all the above information from their safety manual and explained him in detail.

They started brain storming as what could have happened. Generally, CO_2 should be discharged through the horn only. Horn acted like expander to kill the pressure and increase the volume of discharge. Paul's engineer expressed a doubt.

Engineer," Sir, we have to find out whether the cylinder had the hose and horn assembly, when it was flying".

Paul, "Yes. It is a good question. Please find out the answer."

He went out to the shop floor, caught hold of the welder who was near the accident. He recollected the incident and told that the cylinder did not have horn or hose when it was flying.

Paul," What was the starting point of the Rocket?" Welder pointed a machine and told "Bottom of this Machine".

Welder, "If I remember correct, I took the extinguisher from the wall hook and kept it near me on the machine for easy use"

Paul, "So, shall I conclude that the extinguisher fell from the place where you kept it on the machine, due to vibration. Upon falling on the floor, horn and hose assembly got detached."

After a pause, that welder accepted his statement and felt sorry for his mistake. Paul said "Don't worry. Accidents do happen. If we do not learn a lesson or two from them, then only it is wrong. Paul and his engineer went to Roshan's cabin and explained the whole incident. Welder had removed the extinguisher from the wall hook and kept it on a machine which was near to him. Due to vibration, extinguisher fell from the machine and the horn assembly got detached. Liquid CO_2 started gushing out from a small nozzle, leading to jet action. Because of velocity, the extinguisher started flying like a rocket. After making a few rounds, CO_2 got exhausted and the rocket fell down. Roshan and his engineers were fully ir explanation.

Roshan," Now we are clear about the root cause of this accident".

Roshan took them to the Managing Director. Again, Paul explained him the whole thing. He was convinced. Along with his quality engineers, he asked Paul to prepare a report.

Paul, "Sir, it is important to add this point in your safety manual. CO_2 extinguisher should be kept in vertical position and should be kept in the wall hook provided for that, use it and keep it in wall hook only.

Paul thanked all the workmen without whose openness in narrating the true description of the incident, he would not have found out the root cause. Roshan thanked them. They left the factory with full satisfaction.

Proper storage of safety equipment while not in use is as important as usage of that equipment.

XVIII

Does machine only emit toxic gas?

Paul was celebrating his 25 years of career as management consultant, serving in the area of quality management. He had visited hundreds of industries in this connection. BOSS appreciated Paul for his sharp brain to crack the root cause of any Issue.

BOSS said, "Paul, you have well spent 25 years as management consultant, mostly with me. I wish you belter start your own management consulting company now" All colleagues clapped and encouraged Paul to swing into action.

BOSS continued, "I am also growing old. My son is not interested in continuing my business as he is a medical doctor. He wants to grow in his line" Paul said, "I see. Why do you worry sir? We are here to support you".

BOSS said, "As long as I am active and kicking, this is okay. I am getting older and I may not work now as actively as I was last year." All became sad. Everybody accepted the reality. Boss"s idea was to hand over the management consultancy wing to Paul at a reasonable price, in view of his loyal and sincere association with him. Boss said, "Of all my business ventures, only this management consulting wing has given me this much name and fame. I have to acknowledge the contribution by Paul".

Tears came in Paul's eyes and boss hugged him. Paul accepted the offer and promised to run the company very efficiently as it was. Paul promised to retain all employees with him. His colleagues heaved a sigh of relief, as their jobs would not vanish. Boss assured Paul to transfer the ownership to him smoothly and help him to get financial support from bank. He said he would continue as an investor, leaving the entire responsibility to Paul. He was in cloud nine and thanked Boss profusely for giving this bonanza.

BOSS suddenly changed his tone and said "There is a last test for you. My friend is struggling to solve a problem in his industry. Several months passed without a solution. I promised him to solve the issue. I have accepted this challenge, with enormous faith in you." Paul assured to put forth all his efforts to resolve the case. He collected the case file from Boss and studied the same deeply.

It was a power plant in northern part of India. Coal reserves are quite good in India. Coal based power plants pollute the atmosphere heavily and always earned the hatred of environmentalists. Government was in a fix. On one hand, enormous coal reserves were there which could solve the country's huge power demand. On the other hand, environmentalists oppose setting up of coal-based power plants in view of pollution problems. Indian government was frantically looking for clean energy technologies to get over this problem. One company in North India had put up a coal gasification plant and a power plant using coal gas as fuel. This plant had been supported by government under its clean coal initiative. It was a small power plant with 5MW electrical output. Upon successful operation, they planned to scale it to 100MW.Plant produced gas from coal called Blue Water Gas (BWG) which was rich in carbon monoxide. This gas was getting generated in the gas plant. Some quantity of gas was being stored in gas holder as buffer. If there was a small interruption in gas plant operation, gas holder supplied the gas for continues operation of Power plant.

From the gas plant, pipelines carried the gas to the boiler. It had several gas burners to burn the gas at various elevations. Water evaporated into steam in the boiler, which drove the turbine. Generator coupled with the turbine produced electrical power. The pipeline carrying gas had been painted in blue color to indicate Blue Water Gas pipe. This gas contained good amount of carbon monoxide, which was very dangerous to breath. It could not be seen or smelt. Breathing carbon monoxide could cause headache, dizziness, vomiting and nausea. If its levels were to be high enough, one might become unconscious and even die. Low levels inhaled for a long time or high level inhaled even for a short time, could pose serious health hazards. Carbon monoxide was a clean fuel with reasonable calorific value. Hence clean coal initiatives allowed carbon monoxide fired boilers. Adequate care was needed to be taken to see that not even a small leak could be allowed in the pipelines, flanges etc. There were expert mechanical engineers who would design such arrangements. Instrumentation engineers would provide carbon monoxide leakage monitors which sense its concentration in the air. Normally alarm used to be set when the carbon monoxide concentration exceeded 20 parts per million. After the alarm, industrial man coolers (Fans) would be automatically switched on to diffuse the concentration. Fans would be made off, once the concentration came below 20 parts per million.

As per the case file, this plant had installed several carbon monoxide monitors along its pipelines, from gas plant to boiler area. Gas plant also had several such monitors. Paul felt the protection arrangement was quite adequate.

BOSS said, "Paul, I know by this time, you would have studied the case file." Paul said, "yes sir. Tell me what is the issue?"

BOSS explained, "Several times, they get leak alarm for a few minutes, especially in the night. Though man coolers operate and reduce the concentration, operators are afraid of any health hazards, as it is happening often".

Paul said, "Obviously. Nobody will dare to stand near a bomb knowingly for a long period" Boss said, "They are unable to find out the reason. So, they have given me a contract to resolve this. I am passing it on to you, my boy."

Paul readily agreed to plunge into the case and resolved to unfold the mystery. After a few days, Paul left for the plant, along with one colleague. Plant Director welcomed Paul and his colleague. Within a short time, he arranged a meeting with his engineers and they explained the entire episode. Every now and then, the maintenance staff used to check the pipe leakage with soap solution. (It was a standard practice in process industries to detect leaks in pipelines especially in areas like flanges, bends and fittings. They would apply soap solution with a brush on the suspected areas. Should there be a leak, soap bubbles would show up in that place). No leak was detected. Somebody suggested to go in for ultrasonic leak detector to scan the pipe line for any leak. When gas escapes from a pressured pipe line through a small hole, it generates a sound in the range of 25KHZ to 10 MHZ, well above the frequencies the human ear is sensitive to but easily identifiable by ultrasonic sensors.

Pipelines were scanned with ultrasonic monitor. It would detect even a minute leak from the sound it generated. Even ultrasonic monitor did not detect any leak. Engineers felt that there was no leak absolutely, as confirmed by soap solution test and ultrasonic monitor test. Engineers tested the carbon monoxide monitors and recalibrated several times. They could not find any fault, anywhere. They interchanged the carbon monoxide monitor kept near the area where leak was reported with new ones, a few times. Monitors were okay. Engineers were upset that they could not find out the origin of the leak, reported at least once in every night shift. Though engineers wrote it off, plant operators pestered them to locate the cause and set right as they were terribly afraid of inhaling carbon monoxide. Matter went to their home fronts also. Wives of some operators came to the plant and complained with the HR department about the health risks of their husbands. One HR Manager also participated in the meeting. Paul carefully observed all the deliberations. His assistant noted down all the points.

Plant Manager took Paul and his engineer around the plant, specifically gas plant and the gas pipelines. Paul asked plant personnel to do soap solution test in the pipeline where leak reported. Result was perfect and no leak observed. He asked them to bring the ultrasound monitor and measure the leak.There was no leak. By then, day shift closed and the day shift operators started leaving the plant. Paul felt tired and both left for the hotel.

In the hotel, after dinner, Paul and his engineer discussed the whole episode threadbare. They started brain storming. Three hours passed but could not draw any conclusion. They decided to see the video camera footage of the camera near the leakage spot, just to grope in the dark. Next day, Paul asked plant engineer to play the video camera footage, half an hour before and after leakage time. Nothing untoward could be found, excepting one person moving around the area.

Plant Manager said that he was one of the maintenance technicians and moving around as part of his job. Paul then asked for camera footage for the past seven days around the leak time. Paul and his engineer found that the same man is moving around the area. He was just walking in that area, not seen for a few minutes and again seen walking back and leaving the area. Almost same thing was seen in every night. They observed the footage again and again. That man was not doing anything wrong. He did not touch any pipeline or the carbon monoxide monitor. No sabotage suspected. Paul had an intuition that he was nearing the cause. They decided to follow that man during night shift.

Paul and his engineer came to the plant at 10PM. Without anybody's notice, they were subtly present in the leakage reported area. Around 12.30 AM, that man came to that area. They were keenly watching him with a lot of excitement, as if they were watching a James Bond thriller. There was a dark area behind a cupboard. They saw him going to that area. They followed him without his knowledge and started observing his action from a distance. He just took out a Cigar, (Native form of cigarette, which is very strong) lighted it, enjoyed it for about 5 minutes and left that area. They observed the smoke from cigar was going to the carbon monoxide monitor which had a small suction action. They observed that the leak alarm started when he lighted the cigar and vanished a few minutes after he left the place. Paul told his engineer that he would confirm the action of carbon monoxide monitor with smoke from cigar in the laboratory.

Next day, Paul's engineer purchased two cigars from the hotel shop and brought to the laboratory. Cigar was lighted up and kept near a carbon monoxide monitor. To everybody's surprise, monitor sucked the smoke and showed the reading as 25 parts per million which was more than the alarm value set in the monitor. It was shocking to know that cigar emitted so much carbon monoxide! What would happen to the poor man's lungs? Now Paul could conclude the root cause.

Plant manager arranged a meeting with his engineers. Plant Director also attended. Paul explained the reason behind carbon monoxide leak alarm. The concerned person had a habit of smoking cigar during night shift. As smoking was prohibited inside the plant, he found out a hideout behind a cupboard and started smoking every night. Cigar emitted carbon monoxide which was sucked into the monitor kept in that vicinity. As the level was more, alarm got initiated and the man cooler operated. The concerned person did not suspect that it was his cigar that triggered the alarm. He thought it was gas leak and left the place immediately. Plant manager called that person to the meeting hall and scolded him for his misconduct, which resulted in panic in the plant. He understood the reason slowly and apologized for his mistake. He suspended him for one week immediately. Though Paul pleaded to excuse him, plant manager did not budge. Paul asked the Plant Director to allot a room where such people could smoke. Plant Director assured the group to conduct suitable classes to educate on the evils of smoking and make them quit smoking.

Paul and his engineer left the plant with a sense of satisfaction. Paul explained the entire episode to Boss. He was very happy that Paul could precisely locate the root cause in a short time. Plant Director called Boss and appreciated the team for their brilliant work. Plant Director felt very much relieved.

When we suspect a machine to emit carbon monoxide, why not suspect a man? This is a new learning. After all, life is full of learning.

About The Author

Dr.R.Jayapal obtained his B.E(Hons) from Annamalai university, M.Tech from Indian Institute of Technology/Madras and Ph.D from National Institute of Technology, Trichy, all from India. He has around four decades of experience in Bharat Heavy Electricals Limited (BHEL), Cethar Limited and Sri Ramakrishna Institute of Technology, India. He is recipient of NRDC republic day award instituted by Government of India for meritorious inventions and BHEL's highest award, Excel Award for creativity and innovation. He holds four patents. His circuit designs have appeared in popular electronics magazines like EDN, Electronics Design published from USA. He has served as a visiting specialist in University of Tennessee Space institute, Tullahoma, USA. Presently he is a Freelance Automation Consultant and Expert committee member of Startup & Incubation center, NIT, Trichy, India.

NOTES